Math At Your Fingertips

by

Janice Judd Blanchard

and

John D. Jenott

DORRANCE PUBLISHING CO., INC.
PITTSBURGH, PENNSYLVANIA 15222

ISBN # 0-8059-3435-9
Printed in the United States of America

First Printing

To our readers:
All words and thoughts by John D. Jenott appear in italics.

—Janice Judd Blanchard

On the cover:
Famous mathematicians: 1)Kepler, 2)Archimedes, 3) Gauss, 4) Pythagoras, 5) Newton, 6) Ptolemy, 7) Galileo.

TABLE OF CONTENTS

FOREWORD

by John D. Jenott

"Oh, please," Brer Rabbit pleaded, "please, don't throw me into the mathematical patch, Brer Fox. Throw me into the briar patch instead."

Brer Fox thought about it: "If I throws him into the mathematical patch, it won't hurt him, and he might learn something valuable. But I can tell by the way the little scoundrel squeals that if I throws him into the briar patch, he'll get a good scratchin' an' he'll never be worth a doodly-doo."

So he threw Brer Rabbit into the briar patch.

Brer Rabbit went right through the briar patch with nary a scratch, and he scampered away laughing. But Brer Fox was right, that rabbit never was worth a doodly-doo.

* * * * *

I was a Brer Rabbit in school, always avoiding the mathematical patch. After all, I was going to be an artist, and what would an artist need with math? Well, enter the real world: there was proportioning pictures, there were picas and points, bookkeeping, charts, graphs, percentages, clients to bill, and, of course, ever-present taxes to be calculated. All this, and I didn't know doodly-doo. So it was back to math and learning what I should have learned in the mathematical patch at school. If you are having similar problems, this book should help—it helped me.

—JDJ

PREFACE

Math At Your Fingertips is not intended to be complete or in depth in any area. Anything not covered can be filled in by using encyclopedias, dictionaries, almanacs, and math books. This book was meant to be a practical, entertaining, simplistic quick-reference book for adults. Math facts that I found interesting, words found in literature, and subjects to stimulate the imagination and to promote further research have been included. The cartoons are intended to be illustrative explanations of the math principles involved.

Dedicated to my husband, Russ Blanchard.

ACKNOWLEDGEMENTS

For their contributions, support, and inspiration:

Dr. David L. Judd, Senior Lecturer in Physics, University of California, at Berkeley and Senior Research Physicist, Lawrence Berkeley Laboratory

Dr. Rene de Vogelaere, Professor of Mathematics, University of California, at Berkeley

Don Larkin, Marilyn Larkin, and Patrick Larkin

My brother, Robert Judd

Merle Budelman Emerson

Lawrence H. Moe, organist and retired Professor of Music, University of California, at Berkeley

Wayne Carlson

My husband, Russ Blanchard, who encouraged me to pursue this project before he died in 1989

Bob Mauser who inspired me to pick up in 1990 where I'd left off in 1989

John Don Jenott, the artist, for being on the same wavelength

THE CONSTRUCTOR

Janice Judd Blanchard grew up in Berkeley, California, and attended the University of California as a music major. She taught for thirty-two years in California secondary schools, mostly in her major subject. The start of the 1970s saw a change in Berkeley Schools curriculum so she ended her career teaching mathematics. She was really a greenhorn because high school algebra and geometry and a desire to succeed were her only qualifications. This necessitated a concentrated math study during the summer vacation before embarking on presenting the just-introduced New Math. Some of the materials collected during this time are the basis for *Math At Your Fingertips*.

THE ARTIST

John Don Jenott was a ninth and tenth grade English class pupil of Janice Judd Blanchard during her first two years of teaching, in Fort Jones, Siskiyou County, California, during World War II, at which time she was faculty advisor for the school paper *The Fort Crier*. Don Bunker, as he was then known, proved to be an extraordinarily talented cartoonist and poetry contributor for the publication, signing his work "D. Bunk." He went on to art school and became a major illustrator for magazines. Jenott was, of course, the first and only choice as illustrator for this book.

Jenott speaks for himself in the following: *Somewhere around my fifth year I drew a picture of Popeye. Rave reviews from my family told me I was on to something, and I never stopped drawing from that day forward. It was an unnecessarily long circuitous route that I took to becoming an illustrator because, being a backwood-mountain lad, I had no idea you could actually make a living at it. Before finally getting into the commercial art world, I sailed in the merchant marine; did a stint in the Air Force; stereotyped on a newspaper; sold cookware; operated a gold mining dredge; worked for a city gas company, a lumber mill, and the county road department; fought fires for the U.S. Forest Service; and functioned as a draftsman. I had, altogether, a sweaty time of it. After I finally got my act together and went to art school, I became art director for several advertising agencies, ran my own art studio, taught exhibitry and design at a job corps camp, supervised the graphic division of the California Region of the U.S. Forest Service, and finally eleven years ago, went off on my own to freelance. I have never looked back and now work in my studio at home in the mountains of far Northern California.*

—JDJ

LINEAR MEASUREMENTS FROM OUR PAST

For centuries, measurements, you might say, were not exactly standard or precise. Some examples of the way business was done long ago are listed here.

- A cubit (Egypt) was from the elbow to the tip of the middle finger (18 to 19 inches). The Old Testament says that Noah's ark was 300 cubits long, 50 cubits wide, and 30 cubits high.
- A span (Egypt) was from the tip of the thumb to the end of the little finger with the hand spread out.
- A palm (Egypt) was across four fingers or about three inches.
- A digit (Egypt) was the breadth at the middle of the middle finger (1/24 of a cubit).

- A rod (16th C.) was the length of the left foot of 16 men in church one Sunday.
- The yard, decreed by King Henry I of England, was the distance from the tip of his nose to the end of this thumb, or to the tip of his outstretched middle finger.

- The inch, the length of three barleycorns, was invented by the Romans. It divided the foot into twelve parts.
- A fathom (Egypt) was the length of outstretched arms, or about six feet.

NUMBERS WITH SPECIAL MEANINGS

Many cultures and individuals have attributed magical, mystical, or symbolic meanings to certain numbers, or patterns of numbers—some of these meanings involving superstition. The "1", a vertical stroke, represents a digit, but some also regard it as the stroke of Creation. Examples of groups of three are numerous in mythology, religion, and the arts. Pythagoreans called "3" the perfect number. Pythagoreans also gave mystical meanings to numbers, as being good or evil.

There are many references to "7" in literature. One good example is in the book of Revelation in the Bible. The Chaldeans regarded the "7" as a mystic or sacred number, and the Pythagoreans considered it lucky. It is also tied in with astrology. The 7th son of the 7th son was considered notable and was supposed to possess prophetic powers. There's a reference to this in the Fred Astaire/Ginger Rogers film *The Gay Divorcee*.

The number "13" is couched in superstition and is generally thought to be unlucky. Many hotels have no numbered 13th floor. Some attribute the superstition to the Last Supper, and others to the Norse myth with 13 at the table. In the latter, at a banquet in Valhalla, Loki, the spirit of evil, intruded making 13, and Balder was slain.

On the radio and television, Jack Benny made 39 a humorous number, in reference to his age.

The number "40" was widely used in the past to denote "many" so you'll find numerous examples in your reading. The Bible has references to "40", including the flood, and you're probably familiar with the story of Ali Baba and the Forty Thieves. William S. Gilbert had "40 daughters of the major-general" in the cast of 19th century *The Pirates of Penzance*, a satirical comment on past usage of the number, and on the general's prowess.

England does not allow the use of "666" on license plates because some people associate this number with the devil. It's the number denoting the beast in Revelation 13:18 given as "six hundred three score and six" in the King James version and as "666" in the Jerusalem Bible. The California Department of Motor Vehicles distributes a license plate with the number "666", but it can be exchanged at no charge.

German composer Johann Sebastian Bach (1685-1750) was interested in numerology, including the mystical quality of the number "3". He often used numerical games in his music and held mathematics and its relationship to music in high regard. He attached numbers to his name (e.g., B A C H = 14) and used this in his phrasing.

Pythagoras, a 6th century Greek mathematician, founded a secret society which regarded numbers as mystical, constituting the nature of things, or "The Music of the Spheres."

These are but a few examples of numbers which have stirred mankind's imagination and influenced daily life and thinking.

NUMBERS AND COUNTING, FROM THE PAST

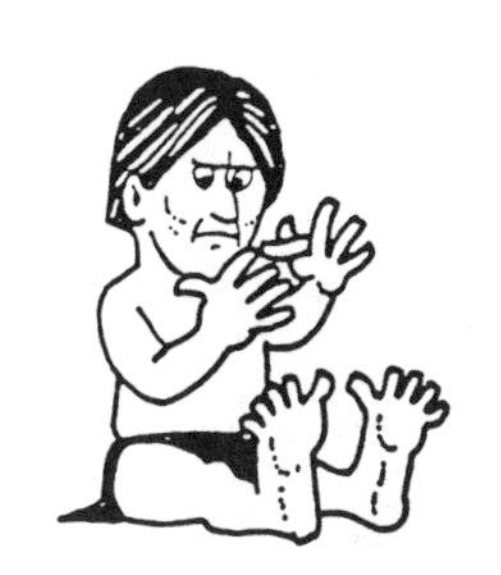

Multiples of 20 resulted from counting fingers and toes among the Mayans, the American Indians, and the Basques. We still have a score which equals 20. Lincoln's Gettysburg Address begins "Fourscore and seven years ago...", denoting 1863 (the year of the address) minus 87, which gives us 1776 because a score equals 20.

Babylonians (Chaldeans) and Persians preferred multiples of 60. This is still found in our time (minutes, seconds) and in angle (360° circle) divisions.

Calculation in Roman numerals is cumbersome so was eagerly replaced when Arabic (Hindu) numerals were introduced, with the zero giving us our decimal system.

Multiplication using Roman numerals:

XX	XVII
XI	XVI
CC (or 20 x 11 = 220)	CLXX (or 17 x 16 = 272)
XX	L X VV
CCXX	X V II
	CLLXXXXVVVII
	CC L VII

This should make us appreciate the decimal system and our method of doing arithmetic!

NUMERATION SYSTEMS

Base 10 (Decimal) which we use in our arithmetic is the superior system, with zero and with place value. It is based on the powers of 10. The number 42 means (4 x 10)+(2 x 1), with the 4 in the 10's place and the 2 in the 1's place.

Base 12 (Duodecimal) is divisible by 1, 2, 3, 4, and 6 so is considered a superior system by many. There is a Duodecimal Society of America in New York. We have the dozen, 12 months, and 12 inches in the foot.

Base 2 (Binary) is very useful today, especially in computers and in space science, with its two positions, open and closed. It is said to have been discovered by the Chinese 4000 years ago, then lost for centuries.

Places are from right to left:

000 = 0
001 = 1
010 = 2
011 = 3
100 = 4
101 = 5
110 = 6
111 = 7

1	1	1	1	0
2^4	2^3	2^2	2^1	(1's)

$16+8 + 4+2 + 0 = 30$

Only the digits 0 and 1 are used. Each place value is multiplied by the power of 2. The speed of transmitting or recording the digits makes up for the large number of digits. A space capsule is racing into space with a binary code message which we hope might be deciphered by intelligent aliens.
Any integer greater than 1 can be used as the base of a system. You might like to work out a base 4 or base 5 system.

ZERO

("Nothing", Cipher, Empty number, Placeholder)

The zero was probably passed on to the West by the Arabs (8th-11th centuries A.D.) who had obtained it from the Hindus. Its introduction gave us our base-10 system of arithmetic and made large number calculation in mathematics and science possible. It also caused the demise of Roman numeral calculation.

The Mayans had a zero symbol, but it wasn't used for calculations. They didn't know what to do with it. The Chinese had no symbol, but apparently an understanding of the zero was implied.

Absolute zero is used to indicate the theoretically lowest possible temperature.

It's helpful to remember that:

$1 + 0 = 1$ Zero when added to a number leaves the number unchanged.

$1 \times 0 = 0$ Any number multiplied by zero equals zero.

$0 \div 1 = 0$ Zero divided by any number is still zero.

The zero is between -1 and +1 on the scale of integers.

<—-1——0——1+—>

CASTING OUT OF NINES

by John D. Jenott, who was once baffled by the mathematical patch

Why throw away perfectly good 9s? Well it seems it was a pretty foolproof way to check your addition before the advent of calculators. You can always add your columns of figures down and then back up to check them, but, the mind plays funny tricks, and if you added 6 and 7 wrong on the way down, then you may very well add them the same on the way up. I don't know who came up with this—probably lost in history now—but I would bet he or she was a bookkeeper.

And why the mysterious 9? I don't know, but it's rather fascinating, and here's the way it works:

Add your columns		*Now throw out the nines and figures that add up to nine*		*Now add across what's left*
3695	⟶	3695	⟶	5
+8142		+8142	⟶	+6
11837		11837	⟶	(11)

same figure so your addition was correct

$1+3+7 = (11)$ ⟵

—JDJ

BATTING AVERAGES IN BASEBALL

If you say a baseball player hit .211, it means he got 211 hits out of 1000, getting on base with no error being called. A batting average of .300 means he got a hit 30% of the time.

A .340 batting average is very good, .400 is phenomenal, and Ted Williams even hit a .406 in his day.

GOLF HANDICAPS

The golf handicap is an equalizer, making it possible for mediocre golfers to play with better golfers and make a decent showing. Determining golf handicaps is so mathematically complicated a computer is now used. In the past a slide rule was used to work out the formula. An example of how the 18-hole handicap is determined is given here. The best ten of the player's most recent twenty scores (e.g., 100, 100, 101, 101, 102, 102, 103, 103, 104, 104,) are averaged. The average 102 is compared to the USGA course rating for the particular golf course (e.g., 67). Thus 102 - 67 = 35. The 35 is multiplied for some reason by 96% (.96). Then 35 x .96 = 33.6. The index number of 33.6 is checked on the golf course's slope rating chart (e.g., slope rating of 120), and we find a handicap of 36. The handicap is then subtracted from the gross score for a net or adjusted score (e.g., 106 - 36 = 70). A handicap of over 40 would be cause to exclude a player from competition on some golf courses. This number varies with the course and the tournament. Needless to say, tournament scratch golf played by pros has no handicaps.

ROMAN NUMERALS

! = I (finger)	II (2), III (3), IIII or IV (4)
5 = V (open hand)	VI (6), VII (7), VIII (8), IX (9)
10 = X (2 Vs)	XX (20), XXX (30), XL (40)
50 = L	LX (60), LXX (70), LXXX (80), XC (90)
100 = C (centum)	CC (200), CCC (300), CD (400)
500 = D	DC (600), DCC (700), DCCC (800), CM (900)
1000 = M (mille)	MM (2000)
5000 = $\overline{V}$	(The vinculum line which multiplies by 1000 was from the Middle Ages.)
1,000,000 = $\overline{M}$	

Rules for the order of the letters:

1) normal order: MDCLXVI = add (MD = 1000 + 500 = 1500)
2) out of order: IX = subtract 1 from 10 = 9

Calculation (see Numbers and Counting from the Past)

Rome had a strong influence in commerce, science, etc. for nearly 2000 years. Roman numerals are still commonly found, especially to indicate the year of printing of a book or movie, on building inscriptions in stone, and on some clock faces.

THUNDER AND LIGHTNING

Most of you have experienced electrical storms and wondered how far away the action was, or whether it was moving toward your area or away from it. An easy way to approximate the distance is to start a count as soon as a flash of lightning is seen. When the count is high enough, you can come out from under the bed.

To determine the distance of a storm from your location, divide the seconds between the flash and the sound (Count "one thousand and one, one thousand and two, one thousand and three, one thousand and four, one thousand and five...") by five for miles, or by 5.5 for nautical miles. Don't take refuge under a tree. It could be dangerous.

Light travels about 186,000 miles/second so you see the lightning almost immediately. Sound travels about 750 miles/hour, or about one mile in five seconds.

GAUSS SOLUTION

In 1801, at the age of ten, Karl Friedrich Gauss did the following problem in two minutes:

Add the numbers 1 to 100

Solution:

$$\begin{array}{llll} 1 + & 2 + & 3 & \ldots\ldots\ldots\ldots + 100 \\ 100 + & 99 + & 98 & \ldots\ldots\ldots\ldots + \;\;1 \\ \hline 101 + & 101 + & 101 & \ldots\ldots\ldots\ldots + 101 \end{array} \begin{array}{l} = 100\text{x}101 = 10100 \\ \swarrow \\ = \frac{10100}{2} = 5050 \end{array}$$

↑

(intervening numbers included)

Gauss lived up to his early promise and became one of the greatest mathematicians of all time.

Karl Friedrich Gauss (1777-1855)

Reading about Herr Gauss can damage a person's self-esteem. This guy did so darn much it just makes you feel as though you have frittered away your life. By the time he was twenty he had accomplished more than the average mere mortal does in twice, nay, thrice that many years.

When he was eighteen he entered the University of Göttingen. He took to it and spent the rest of his life there, turning down numerous offers from other universities.

It was also at eighteen that he formulated a discovery that had evaded geometers since the time of Euclid, and that was the division of a circle into seventeen equal parts. This is fascinating, and I don't have the foggiest of what the practical use of this is, but it would probably be of great interest to pizza cutters.

In 1801 an asteroid, Ceres, made its appearance. Leading astronomers made up to forty observations of it but were unable to calculate its orbit. Gauss made three observations and developed a technique precisely calculating its orbit. The method is still in use today and, with only a few modifications, has been adapted for computers. But of course, Gauss was a mature twenty-four at the time.

Gauss was a hands-on workaholic (and probably an insomniac) and really did not care for teaching. He would only take a very few students at a time, guided by his favorite saying, "Few, but ripe." You have to be a hard person not to love that saying.

The rest of this book could easily be taken up by his work, inventions, and discoveries in the fields of mathematics, astronomy, mechanics, electricity, magnetism, optics, and statistics; but this book is not about Gauss although his work pervades the whole of it.

There is a portrait of Gauss in his later years. He looks out at you in rumpled clothing with tired eyes. I wonder why.

—JDJ

PROBABILITY AND ODDS

(Chance and the Likelihood of Occurrence)

Our lives are affected daily by probability or odds, not only in familiar gambling pursuits such as slot machines, sweepstakes, and lotteries, but also in insurance policies for health, life, and car. The actuarial insurance tables are based on probability. Slot machine odds are regulated and changed by the casinos, and are operated to the advantage of the house. The odds are much better in this type of gambling than in sweepstakes, or in lottery odds, which are millions to one. The California lotto game had lottery odds of 13,983,816 to one at one time.

Probability enters into engineering when statistics are required to take the safety factor into account. When constructing a bridge, for example, the odds of a catastrophe such as an earthquake, are considered, and anything beyond some number of standard deviations is discarded from the plans.

Our language is riddled with such phrases as "chances are" or "what are the odds?"

At the race track betting odds of 5 to 1 mean the odds are that the horse loses.

The probability of losing is 5 in 6

$$\left(\frac{5}{5+1} \text{ or } \frac{5}{6} \right).$$

If your horse wins at these odds, you receive $12 for your $2 bet, your $2 plus $10 winnings from the bookmaker. If your horse wins at 14 to 1 odds, you receive $30, your $2 bet plus $28 winnings from the bookmaker.

PROBABILITY AND ODDS

A 17th century French mathematician Blaise Pascal developed a theory of probability, called Pascal's Triangle, regarding the tossing of two coins.

Each figure is the sum of the two figures just to the left and right of it in the line above, except that the figures at the ends are all ones.

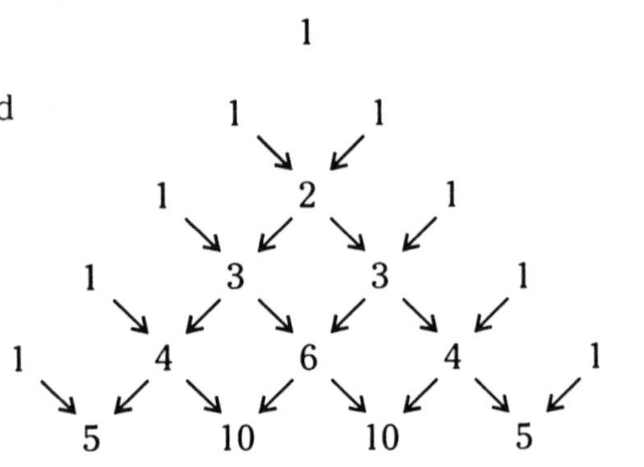

Odds in the toss of two coins:

Line 1: no coin toss

Line 2: one coin toss (1 + 1 = 2)

Odds of a head or tail are equal or 1 in 2 ($1/2$ each)

Line 3: two coin tosses (1 + 2 + 1 = 4)

Odds: 2 heads 1 in 4, 1 head and 1 tail 2 in 4,
2 tails 1 in 4

Line 6: 5 coin tosses (1 + 5 + 10 + 10 + 5 + 1 = 32)

Odds: 5 heads 1 in 32, 4 heads 1 tail 5 in 32,
3 heads 2 tails 10 in 32,
2 heads 3 tails 10 in 32,
1 head 4 tails 5 in 32,
5 tails 1 in 32

So it is with all the lines, *ad infinitum*.

HECK, PASCAL, YOU ALWAYS WIN!

SOUND

The speed of sound depends on the temperature of the air. At 0° Celsius it equals about 1089 feet/second or 742.5 miles/hour. At 20° Celsius it equals about 1130 feet/second or 770 miles/hour.

The sound barrier (speed of sound = Mach 1) was the subject of a tragic, then triumphant 1953 British film *Breaking The Sound Barrier* with actor Ralph Richardson.

MATHEMATICS AND MUSIC

Music is truly a mathematical subject. Tuning, vibrations and frequencies, intervals, scales, and harmonics or partials are manifestations of this. If you've ever watched a piano tuner at work, you've observed this tuner determining beats, by listening, or by using a machine, in order to adjust the pins for satisfactory sounds. This all comes about because of the compromises made necessary in dividing the octave into twelve equal semitones (the chromatic scale) and in tuning the intervals within the octave. Mathematically speaking, it is very compli cated. If you divide 12 into 256 vib./sec., the difference between the C above middle C (512 vib./sec.) and middle C (256 vib./sec.), you get 21.333+. Perhaps you begin to see the difficulties. *The Harvard Dictionary of Music* listed in the bibliography has detailed mathematical working for tuning.

Stringed instruments can make a difference in the enharmonic G$^{\#}$ and A^{b}, with the G$^{\#}$ at 773 vib./sec. and A^{b} at 814 vib./sec., but the piano has to use the same key for both notes. There were actually some 16th, 17th, and 18th century organs which had divided or split keys so that the organist could get flexibility in halftones. The front half of the key played one pipe, and the back another.

You might wonder at pitch conflicts between flexible-pitch strings and fixed-pitch winds in orchestras. There have to be many adjustments by the string and wind players. When a keyboard instrument is used with the orchestra, there have to be adjustments, also.

The modern, music U-shaped tuning fork is usually found at A-440 vib./sec. (U.S. concert pitch) or A-435 vib./sec. (Europe international pitch). This is in the octave above middle C (256 vib./sec.). The oboe A-440 is usually sounded for tuning because of the instrument's more constant pitch reliability. The orchestra pitch "A" has varied in practice, with some groups using a lower pitch, others a higher pitch. The Boston Symphony and other modern orchestras have used A-444 vib./sec. for a more brilliant sound.

In early Greek music, the Pythagorean scale was based on the interval of the pure fifth (equal temperament tuning has a slight lowering of the no-beat perfect fifth to one beat/sec.), and the whole tone which varied but was slightly smaller than today's whole tone. Pythagoras, the 6th century B.C. Greek mathematician discovered numerical ratios of the musical scale. He also founded a secret society which regarded numbers as mystical, constituting the nature of things or "The Music of the Spheres."

Pythagoras

If you think mathematicians lead unexciting lives, consider Pythagoras (580-500 B.C....more or less).

To say this fellow was different is like saying Attila was fractious.

Born on the island of Samos, Greece, he early on took to traveling—and never looked back. For about thirty years he wandered around Egypt, Asia, and India. If his father warned him to beware of those foreigners and their strange ideas he did not listen because he picked up all kinds of notions concerning mysticism, mathematics, and such.

He finally settled down in the ancient city of Croton in southern Italy and quickly picked up a large following of disciples that he organized in a super-secret society.

His main theorem was about numbers. Numbers dominated all. They were the beginning and the end, the essence of all things, things were only the copy of numbers, numbers were themselves the things.... Do you understand this concept? Good. I had a feeling I was alone. Anyway, he would have loved state lotteries.

All was going swimmingly with Pythagoras and his group until they started messing around in local politics. This was a bad move. The local populace, who were mighty suspicious of his secret society, took umbrage at some political ploy the Pythagoreans tried to pull off and proceeded to burn down their building, killing quite a few members of the society.

Pythagoras escaped to another city in southern Italy and set up a new shop. He never attempted to organize another secret society (as they say, "Once burned, etc."), but instead taught and advanced his studies in mathematics, music, architecture, and astronomy. Apparently he was recognized for his achievements in his own time because his tomb was still known to the time of Cicero (106-43 B.C.).

Moral: Avoid politics.

—JDJ

The "mean-tone" system of tuning, which in the 15th through 18th centuries came to be used in Europe, had a slightly lowered fifth (the fifth pulled in a bit) and acoustically perfect thirds. Since music was linear instead of harmonic, and modulation into different keys was not needed, this system proved workable into J.S. Bach's day. Bach pushed a modified tuning of certain intervals in order to fit the system. This "well-tempered" system accommodated the bad keys of the "mean-tone" system. Modulation was possible to "close" keys in "mean-tone", to farther keys in "well-tempered", although some modulations were better than others.

"Equal temperament" tuning replaced "mean-tone" and "well-tempered" tuning in the 19th century out of necessity, to accommodate the harmonic style of composition. With its twelve semitones (chromatic scale), a compromise was made in dividing up the octave, the only pure or perfect interval acoustically with no beats. This compromise was spread over all the tones with tuning determined by logarithms. Now modulation from one key to another was equally possible. This system is still accepted today. Adjustments of the mathematical ratios of intervals are employed by instrumentalists in order to create pleasing sounds. The equal temperament fifth has more beats than the mean-tone fifth. A beat results from the interference of two sound waves of different frequencies (e.g., 435 vib./sec. and 440 vib./sec.). Slow beats (2/sec. to 4/sec.) are acceptable, but sounds can be unpleasant as beats increase.

Human hearing ranges from 20 vib./sec. to 20,000 vib./sec. with, of course, human differences. Below this range is subsonic, above is ultrasonic. The piano keyboard ranges form 40 vib./sec. (or a bit lower) to 4000 vib./sec. (usually). Pianos vary in their number of keys, thus often extend the range. On a pipe organ tuned to A-440 vib./sec. A 32-foot C, 4 octaves below middle C, vibrates at 16 vib./sec., below human hearing. You might wonder about its reason for being. The vibrations below the range of human hearing reinforce partials or harmonics (notes vibrating in sympathy above the fundamental note) which enhance the sound. There are also some 64-foot pipes, one example in the Atlantic City theater organ.

Then, there's the human voice. The vocal cords which produce the sound can vibrate from F below low C (about 40 vib./sec.) which we find in cultivated Russian basses, to the highest recorded soprano C above high C (2048 vib./sec.). At puberty a boy's voice drops in range, sometimes overnight, but usually it's a change over a considerable period of time. When one of the two vocal cords thickens before the other, the result can be a "break" in the voice, a startling jump from low (vibrations on the thickened cord) to high (vibrations on the thinner cord). The yodel can be developed by using the old voice (falsetto) in conjunction with the new lowered voice.

We see that mathematics is an integral part of music, so is it any wonder that so many mathematicians are heavily into music?

ROUNDING OFF

If the number on the end is 5 or more, add 1 to the left. If it's 4 or less, drop it.

Light travels 186,326 miles/second. Rounded off it's 186,000 miles/second.

Rounding off to the tenth: 21 = 20, 16 = 20

Rounding off to the hundredth: 423 = 400, 375 = 400

1.46536554	(Drop the 4)
1.4653655	(Boost to even)
1.465366	(Add 1)
1.46537	(Add 1)
1.4654	(Drop the 4)
1.465	(Chop to even)
1.46	(Add 1)
1.5	(Add 1)
2.0	

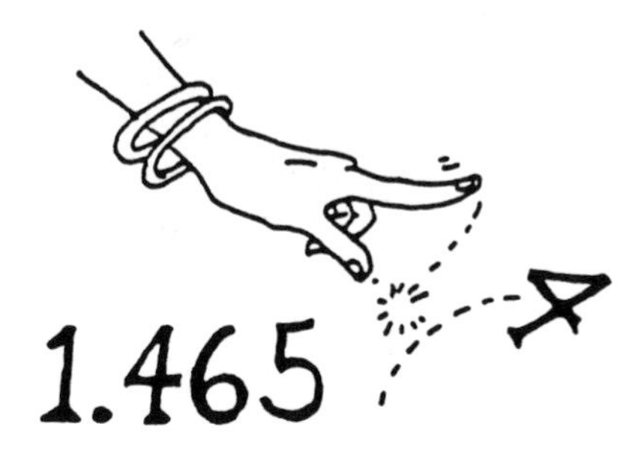

LARGE NUMBERS AND SCIENTIFIC NOTATION

one thousand	1,000 or 10^3 (read 10 to the third power)
one million	1,000,000 or 10^6
one billion	1,000,000,000 or 10^9 (U.S. and France)
one billion	1,000,000,000,000 or 10^{12} (England and Germany)
one trillion	1,000,000,000,000 or 10^{12} (U.S. and France)
one trillion	1,000,000,000,000,000,000 or 10^{18} (England and Germany)

Examples of scientific notation: 3.457×10^5; 1.2×10^3

There is a single digit to the left of the decimal point. The notation is a space saver and an aid to comprehension. You will find reference to the googol, a 1 followed by 100 zeros. It was named by a nine-year-old boy in 1955. A googolplex is defined as a huge number ($10^{10^{100}}$). If someone asks you what the largest number is, tell the person you can always add one to any amount.

It's astounding how small numbers can compound into astronomical numbers, like the national debt. If you double the grains of wheat for each of the 64 squares on a chess board, it gives an amount that would cover the country of India to a depth of several feet.

$2^0 + 2^1 + 2^2 + 2^3 + 2^4 \ldots\ldots\ldots\ldots\ldots + 2^{63}$ = more than
18,000,000,000,000,000,000 = 10^{19} grains

MISCELLANEOUS

To figure your car mileage, divide the miles by the number of gallons used. If you have trouble remembering which number is divided by which number, use a simple example: a car which gets ten miles to a gallon of gas, will get twenty miles from two gallons. Miles divided by the gallons gives you ten miles to the gallon. Use this formula for more difficult calculations.

Position is an important part of our decimal system. The number 1234.56 has:

thousands	hundreds	tens	ones	tenths	hundredths
1	2	3	4	5	6

International time is not only used in the military services, but in travel or in doing business world-wide, it's the one to use.

1:00 A.M. = 0100 (Read "zero or 'oh' one hundred hours")
2:30 A.M. = 0230 hours
12:00 noon = 1200 hours
2:15 P.M. = 1415 hours
12:00 midnight = 2400 hours

Chisenbop (Chisanbop, Chisambop) is a method invented by a Korean mathematician for using the fingers as a calculator.

Numbers twenty-one to ninety-nine are hyphenated.

Abacus

The earliest "calculator" known to history. It has been in use for literally thousands of years. In fact, abacuses were used in prehistoric times before humans could write! That has to be true; it was in an encyclopedia.

The word "abacus" comes from the Semitic word "abg"—try to pronounce that—and it means dust. There seems to be something biblical about that..."dust to dust" etc. The name comes from the earliest forms of the abacus, which were boards covered with fine dust.

Ancient Greek writings assert that Pythagoras taught geometry as well as arithmetic upon the abacus. The Romans used it during the classical period.

It is still very much in use today. In San Francisco's Chinatown you can peer through a merchant's window and see fingers, in a blur, moving the beads up and down. People who use these little beaded wonders are the same brain-type people who have inflicted the computer upon us.

—JDJ

ABACUS

(A bead calculator still in use)

Movable counters and beads were used in ancient times, and calculating was done by sliding these beads. The abacus is still used in Russia, the Orient, and the Middle East. It can be faster than a calculator. I saw it demonstrated by a clerk during my 1960 trip to Russia. Her hands fairly flew over the instrument, pushing the beads up and down.

An example of an abacus and its use:

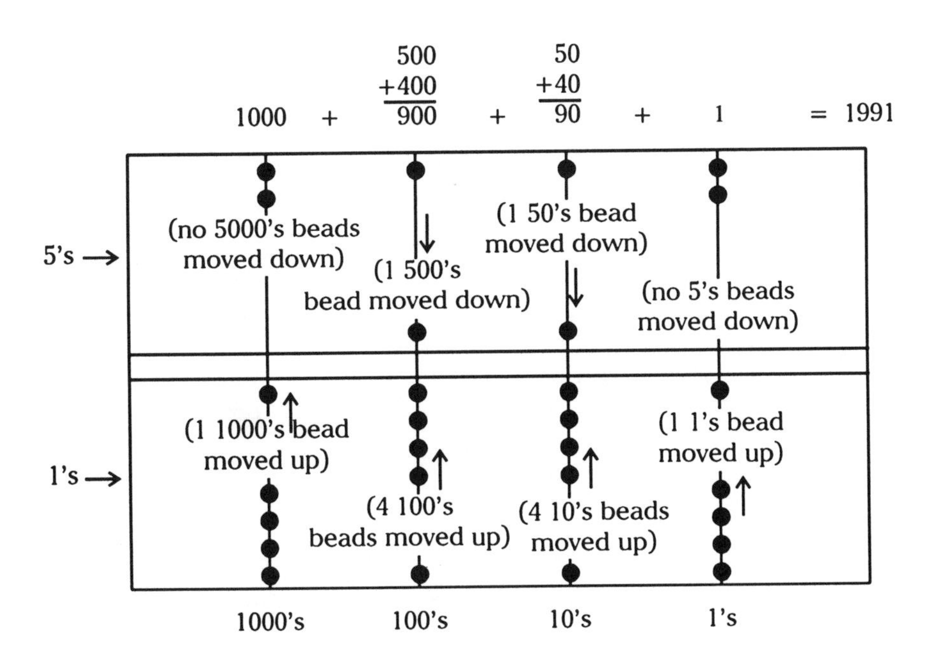

NEW MATH

The New Math was widely used in the schools in the 1960s and 1970s. Terms used for properties of rational numbers were confusing to students, especially those who had difficulty in reading. Satirist Tom Lehrer recorded a definitive commentary on New Math, remarking on the following:

- They teach 2 x 4 = 4 x 2 is commutative, but not that 2 x 4 = 8.
- They teach that 1 + 2 + 3 = 3 + 2 + 1 (associative), but not that 1 + 2 + 3 = 6.
- They teach that 2 (a + b) = 2a + 2b (distributive), but calculations were deemphasized.

Time spent on mathematical principles resulted in a sad lack of calculation skills.

EARTHQUAKE MEASUREMENT

(Recorded on Seismographs)

- The Richter Scale measures the energy at the focus or epicenter.
- The scale is logarithmic, increasing by the powers of ten.
- A 6.0 quake is 10 times the power of a 5.0 quake, for example.
- A 9.0 quake is a million times stronger than a 3.0 quake.
- The Mercali Scale measures damage at the surface.

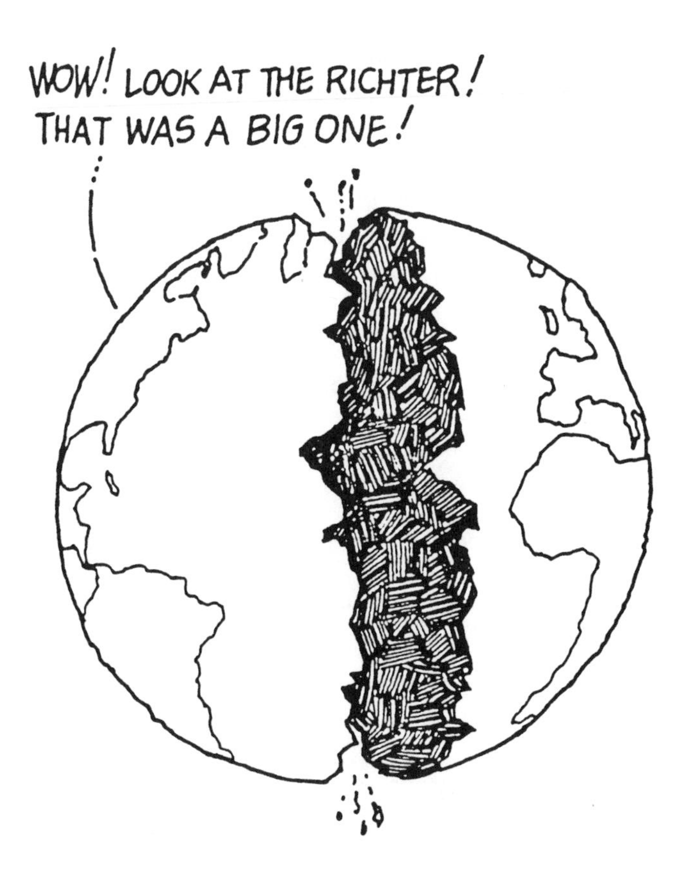

TEMPERATURE

Temperature Measurements (at sea level)

- Fahrenheit (early 18th century, Amsterdam): Water boils at 212°; water freezes at 32°.
- Réaumur (18th century, France): Water boils at 80°; water freezes at 0°.
- Celsius (18th century, Sweden): Water boils at 100°; water freezes at 0°.

"Centigrade" became "celsius" in 1948. I can remember "centigrade" was used in high school chemistry in 1935.

(9/5 Celsius) + 32 = Fahrenheit; (Fahrenheit - 32) x 5/9 = Celsius. We commonly use Fahrenheit thermometers, but it's wise to be acquainted with Celsius for use in science studies and for travel.

Wind Chill (or Windchill) is the measure of the cooling power of air in relation to temperature and wind speed.
20° F. has the same effects as 3° F. when the wind is 10 miles/hour.
20° F. has the same effects as -20° F. when the wind is 35 miles/hour.

Temperature and Air Pressure

Early explorers used the temperature at which water boils to determine altitudes. Water boils at a lower temperature at heights, and cooking takes longer. Generally, water boils at 1°F less per 550 feet of altitude, so at 5500 feet it boils at 202°. In Denver, make a 3-minute boiled egg order, a 4-minute egg.

In a pressure cooker water boils at a higher temperature, with the increased pressure, so cooking is faster.

FRACTION AND DECIMAL EQUIVALENTS

Fraction	Decimal
$\frac{1}{2}$	.5 (read 5 tenths)
$\frac{1}{4}$	.25 (read 25 hundredths)
$\frac{3}{4}$	.75
$\frac{1}{3}$	.333.....
$\frac{2}{3}$	.666.....
$\frac{1}{5}$	.2
$\frac{2}{5}$	.4
$\frac{3}{5}$	.6
$\frac{4}{5}$	.8
$\frac{1}{8}$	.125
$\frac{3}{8}$	.375
$\frac{7}{8}$	.875
$\frac{1}{6}$	.166.....
$\frac{5}{6}$	.833.....

For those not listed, divide a given fraction numerator by the denominator in order to get the decimal equivalent. In reverse read the decimal in terms of a fraction and simplify the fraction. For example:

$$.05 = 5 \text{ hundredths} = \frac{5}{100} = \frac{1}{20}$$

DIVISION TRIVIA

Various methods of doing division problems:

```
      8 3
15 | 1 2 4 5
   - 1 2 0
       4 5
     - 4 5
         0
```

Recording on the side

```
15 | 1 2 4 5 |
   -   9 0 0    60     100's
       3 4 5 |
       3 0 0 |  20      10's
         4 5 |
         4 5 |   3       1's
                83
```

Before the introduction of the calculator, division problems were done by various mechanical means. It's a good idea to know one method of solution in case a calculator is not at hand.

French method: 125 divided by 16

```
          1 2 5      |16
 (16x7)   1 1 2        7.8  answer
            1 3 0
  (16x8)    1 2 8
                2  remainder
```

Why the second fraction is inverted in division:

$$\frac{2}{3} \div \frac{1}{4} \text{ or } \frac{\frac{2}{3}}{\frac{1}{4}}$$

$$\frac{\frac{2}{3} \times \frac{4}{1}}{\frac{1}{4} \times \frac{4}{1}} = \frac{2}{3} \times \frac{4}{1}$$

SQUARE ROOT

If you need the square root of a number, use a table of square roots, a slide rule, or a calculator. Here's how it can be done from scratch.

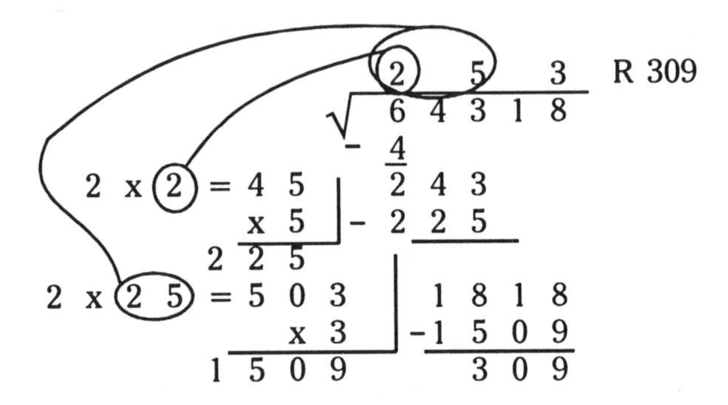

The largest square that will go into 6 is 2 x 2 (4). Subtract and bring down two numbers (43). To the left multiply 2 times the partial answer (2) and add to this (4) a trial divisor (5) which when added to the 40 and multiplied times the 5 will give a number lower than the 243 (225). Repeat the procedure, multiplying 2 times the partial answer (25) and find another trial divisor (3).

Check:

```
      2 5 3
    x 2 5 3
  ---------
  6 4 0 0 9
    + 3 0 9   remainder
  ---------
  6 4 3 1 8
```

A number has two square roots, positive and negative.
e.g., +3 times +3 = 9
-3 times -3 = 9
The square root of 9 is plus or minus 3.

POLYHEDRONS (POLYHEDRA)

A polyhedron is a closed solid with many planes (flat) faces. Each face is a polygon which is a closed plane figure with straight sides and three or more angles. A regular polyhedron ("Platonic Solid") has faces which are congruent (identical regular polygons with equal angles). These were known to the Greeks before the time of Plato. In the 1750s mathematician Leonhard Euler proved that there are five and only five regular convex polyhedrons, in other words a limit was set. The formula as stated by Euler is called Euler's Theorem, or the Descartes-Euler Polyhedral Formula since it was known to mathematician Descartes a century earlier:

$$V - E + F = 2$$

V = the number of vertices
E = the number of edges
F = the number of sides

The Regular Polyhedrons:

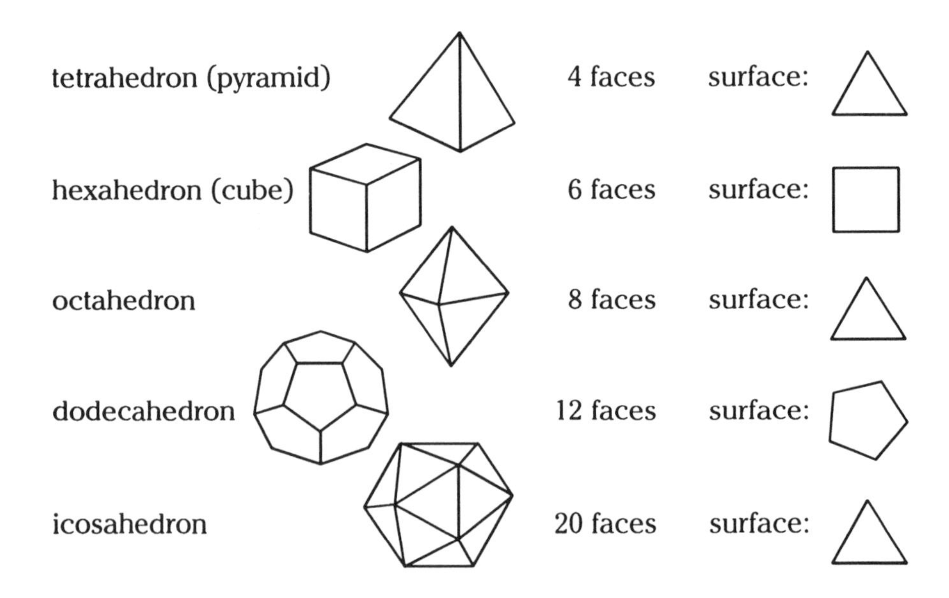

A BOOK OF INTEREST

Flatland, A Romance of Many Dimensions
written by Edwin A. Abbott in 1884,
with illustrations by the author, A Square
(New York: Barnes & Noble, Inc., 1963)

This book regarding dimensions and mathematics is available in paperback. The hero, a square, lives in a pentagonal house with his family, in the 2-dimensional world of Flatland. He gets into great difficulties with his fellow-Flatlanders after he's visited by a sphere from the 3-dimensional world who shows him the wonders of this new concept.

The inhabitants of Flatland are figures from plane geometry: lines, triangles, squares, and polygons from pentagons to approximations of a circle. Their social status depends on the number of sides they have, the highest being the priestly approximation of the circle.

The square tries to explain a sphere cutting through Flatland, seen only as circles increasing and decreasing in diameter. It should stir the imagination of possible fourth or higher dimensions, dimensions which are known only to the mathematician.

THE GOLDEN MEAN

("Golden Ratio", "Divine Proportion", "Golden Section", "Divine Section")

For centuries the aesthetic preference for The Golden Mean proportions has been apparent. The Golden Mean proportions were used by artists and architects, consciously or unconsciously, because it was considered most pleasing to the eye. For some there was even a mystical meaning.

The Great Pyramid of Gizeh (260 B.C.), the Athens Acropolis (5th century, B.C.) with its "Golden Rectangle" proportions, and Le Corbusier's United Nations building in New York City are manifestations of The Golden Mean.

The wonder of the "Golden Rectangle" is found in the ratios which come to the same number after the decimal point.

$$\frac{W}{L} = \frac{L}{W+L} \qquad \text{If } L = 1, W = 0.618034$$

$$\text{If } W = 1, L = 1.618034$$

If you cut off squares successively, you're left with "Golden Rectangles" *ad infinitum*.

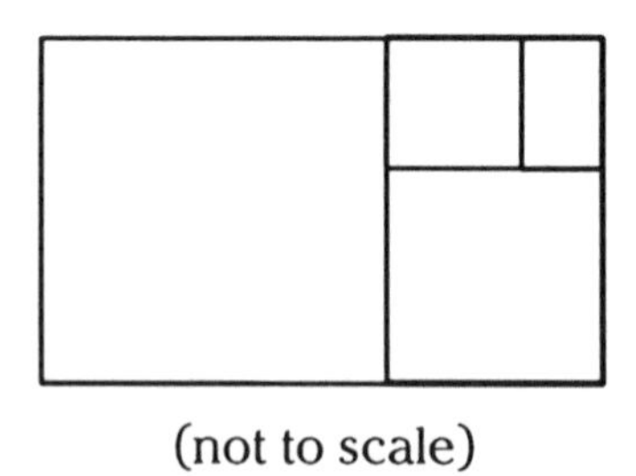

(not to scale)

Artist Leonardo da Vinci (15th-16th century) called it the "divine proportion."

Mathematician Leonardo da Pisa, a.k.a. Fibonacci, who promoted the replacement of Roman numerals with Arabic numerals in the Western World, figured out a sequence of numbers formed by adding consecutive numbers, ratios of which eventually approximate The Golden Mean ratios.

Fibonacci Numbers: 0 1 1 2 3 5 8 13 21 34 55 89 144 etc.

e.g.: 1+2=3, 2+3=5, 3+5=8, etc.

$\frac{13}{21} = 0.619047$ $\frac{21}{13} = 1.615384$

$\frac{233}{377} = 0.618037$ $\frac{377}{233} = 1.618025$

An example of the conscious use of The Golden Mean proportions is found in Béla Bartók's Music For Strings, Percussion, and Celeste, and the placement of the emphasized beats.

The Golden Mean:

When I was in art school I took a course in sho-card design and lettering. Surprisingly the cards always looked best when done on a rectangular format. The instructor never explained that it was the "Golden Rectangle," and I doubt he was even aware of it. I certainly wasn't until I read this.

Incidently I never did become a good lettering artist, but the shape of the cards looked great.

—JDJ

SIGNS

+	add
-	subtract
x or •	multiply
÷	divide
$\sqrt{\ }$	square root
∩	arc
'	foot; minute
"	inch; second
=	is equal to
≠	is not equal to
≐	is approximately equal to
~	is similar to
>	is greater than
≯	is not greater than
<	is less than
≮	is not less than
≥	is greater than or equal to
≱	is not greater than or equal to
≤	is less than or equal to
↔	corresponds to
≅	is congruent to
≡	2 phrases equivalent
$\overline{a+b}$	viniculum: line over several terms
$\overrightarrow{AB}$	vector (point A to point B)
∞	infinity
∴	therefore
%	per cent or percent
( )	parentheses
[ ]	brackets
{ }	braces

Constructions:

For several years I worked as a cadastral draftsman, which means I drew maps from the surveyors' notes.

Every day I used the compass and straightedge and every year I had to get stronger and stronger eyeglasses. I promised myself that someday I would have my own commercial art studio and not allow a compass and straightedge on the premises.

I now have that studio and on nearly every job I use—you guessed it—a compass and straightedge.

—JDJ

CONSTRUCTIONS

Using a compass and a straightedge

Triangle (given 3 sides)

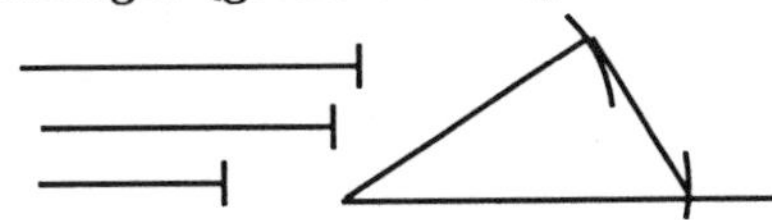

Bisecting an angle

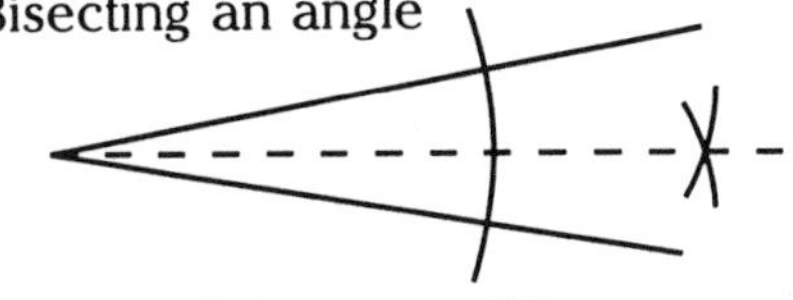

An angle equal to a given angle (vertex and side given)

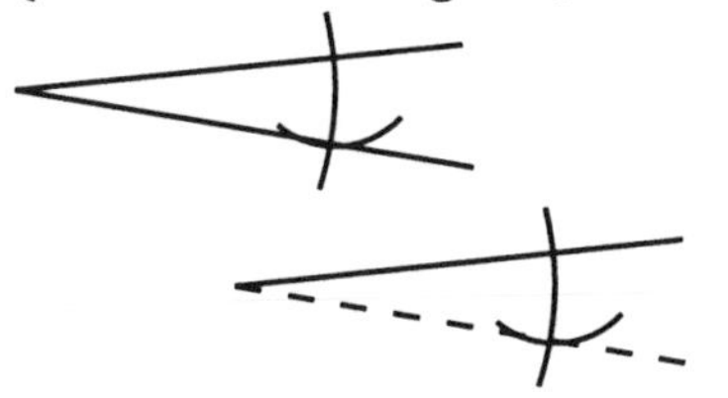

Triangle (given two sides and the included angle)

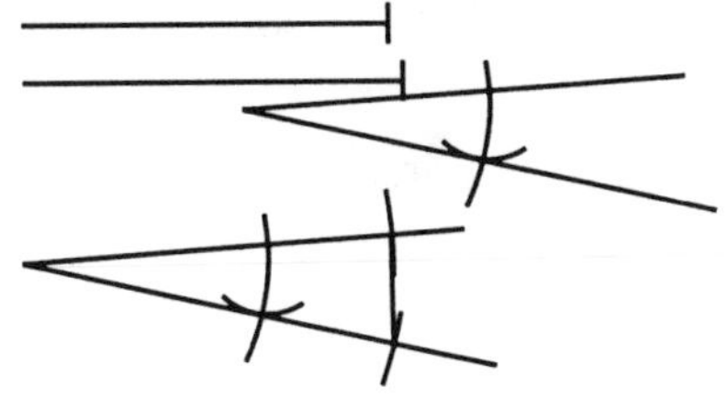

Triangle (given 2 angles and the included side)

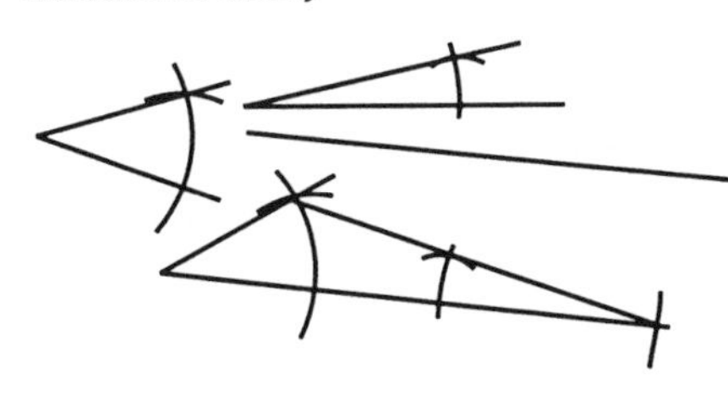

A line parallel to a given line

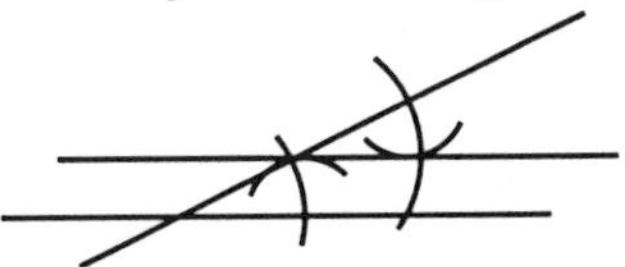

Perpendicular to a line from a given point not on the line

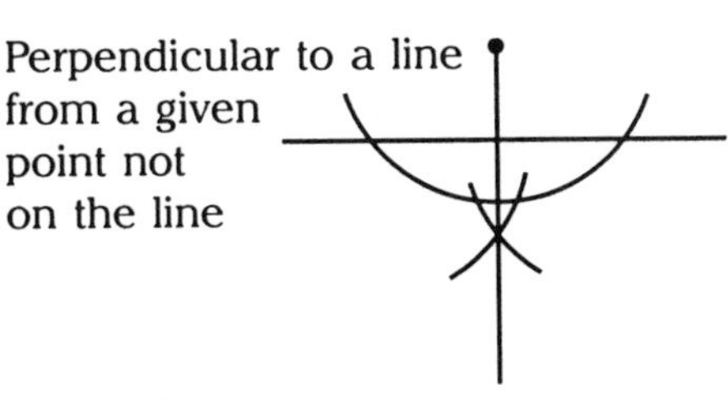

Perpendicular bisector of a line segment

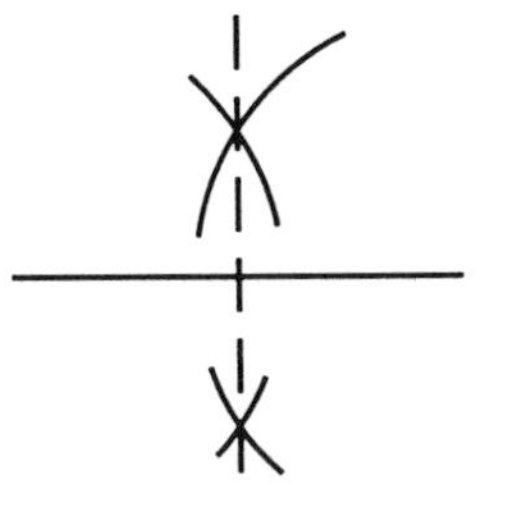

Perpendicular to a line at a point in the line

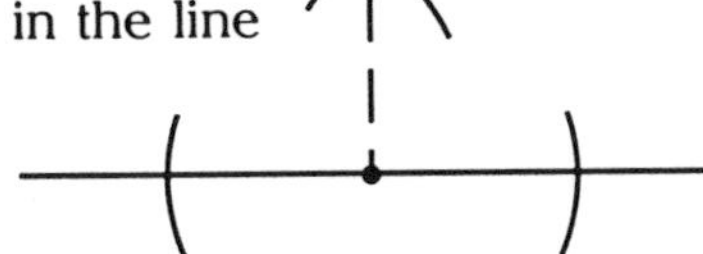

GEOMETRY

(Greek: Earth Measuring)

Many of you can remember your high school "Elementary Plane Geometry" (2-dimensional), with deductive reasoning used to prove that:

- Two lines are perpendicular if they form a right angle with each other.
- Two angles are complementary if their sum is a right angle.
- Two angles are supplementary if their sum is a straight angle.
- Two triangles are congruent if 2 sides and the included angle are equal.
- Two triangles are congruent if 2 angles and the included side are equal.
- Two lines are parallel if they do not intersect even if extended.

etc.

Straightedge, compass, and protractor were the tools. Proof of a theorem was done in a formalized manner with "statements" and "reasons" listed. Euclid, the 4th century B.C. Greek who taught in Alexandria, was the organizer of this plane geometry, concerned with the properties of, and relationships between, points, lines, planes, and figures. Other geometries include 3-dimensional solid, analytic, algebraic, descriptive, differential, projective, non-Euclidian, and affine.

Those who play billiards know that success depends on the use of angles in determining where to hit the ball with the cue.

Geometry:

Roughly translated: Geo means earth and metry (metron) means measure, so it pertains to measurements of the earth. It's from Greek so it is quite okay if you say, "It's all Greek to me."

Euclid (3rd-2nd century B.C.), a Greek mathematician, was big on geometry and created a number of theorems on it. He was a teacher and founder of a school of mathematics in Alexandria. When Ptolemy asked him to show him an easy way to learn geometry, Euclid replied, "Sorry Boss, but there is no royal road to geometry." And that is another rough translation.

—JDJ

GEOMETRY FORMULAS

V = volume; P = perimeter; A= area; h = height; b = base; l = length; w = width; C = circumference; π = pi (3.1416); d = diameter; r = radius; s = side; a = altitude

---

Square: $P = 4s$; $A = s^2$

Rectangle: $P = 2(l + w)$; $A = lw$

Parallelogram and Rhombus: $A = hb$

Triangle: $A = \frac{1}{2} hb$

Trapezoid: $A = \frac{h(b + b')}{2}$

Circle: $C = \pi d$; $A = \pi r^2$

Cone: $V = \frac{1}{3} r^2 h$

Cube: $V = s^3$

Cylinder: $V = \pi r^2 h$

Sphere: $V = \frac{4}{3} \pi r^3$

Sphere surface $= 4\pi r^2$

Pyramid: $V = \frac{1}{3} hb$

Pythagorean theorem for right triangles: $a^2 + b^2 = c^2$ (hypotenuse)

for example: $3^2 + 4^2 = 5^2$

$9 + 16 = 25$

TRIGONOMETRY

Trigonometry is from the Greek and means measurement of triangles. Originally, it had to do with the functions regarding right triangles.

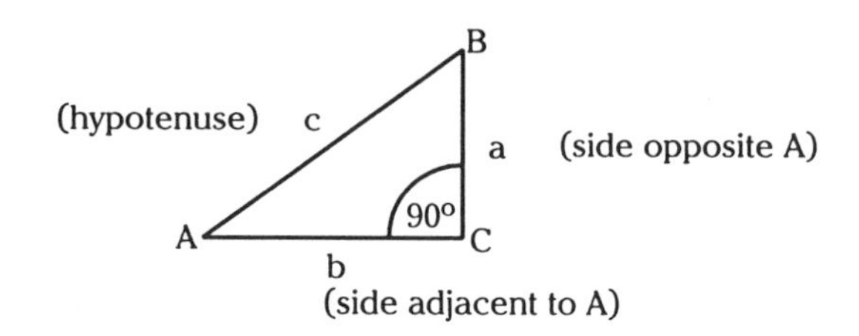

sine of ∠A (sin A)	=	$\frac{a}{c}$	(side opposite) / (hypotenuse)
cosine of ∠A (cos A)	=	$\frac{b}{c}$	(side adjacent) / (hypotenuse)
tangent of ∠A (tan A)	=	$\frac{a}{b}$	(side opposite) / (side adjacent)
cotangent of ∠A (cot A)	=	$\frac{b}{a}$	(side adjacent) / (side opposite)
secant of ∠A (sec A)	=	$\frac{c}{b}$	(hypotenuse) / (side adjacent)
cosecant of ∠A (csc A)	=	$\frac{c}{a}$	(hypotenuse) / (side opposite)

These functions would carry on for angles B and C.

Spherical trigonometry is the study of triangles on the surface of a sphere and is used in navigation, astronomy, and surveying.

MEASURING THE HEIGHT OF A REGULAR PYRAMID

This process uses the properties of an isosceles triangle (the legs opposite equal angles are equal) to measure a height. Sir Arthur Conan Doyle had Sherlock Holmes solve a mystery in *The Musgrave Ritual* by using this principle.

When the shadow "a" equals the height of the pole, the sun is 45° above the horizon. Then measure C to X (the pyramid shadow). BX + XC = the height of the pyramid because triangle ABC is an isosceles triangle.

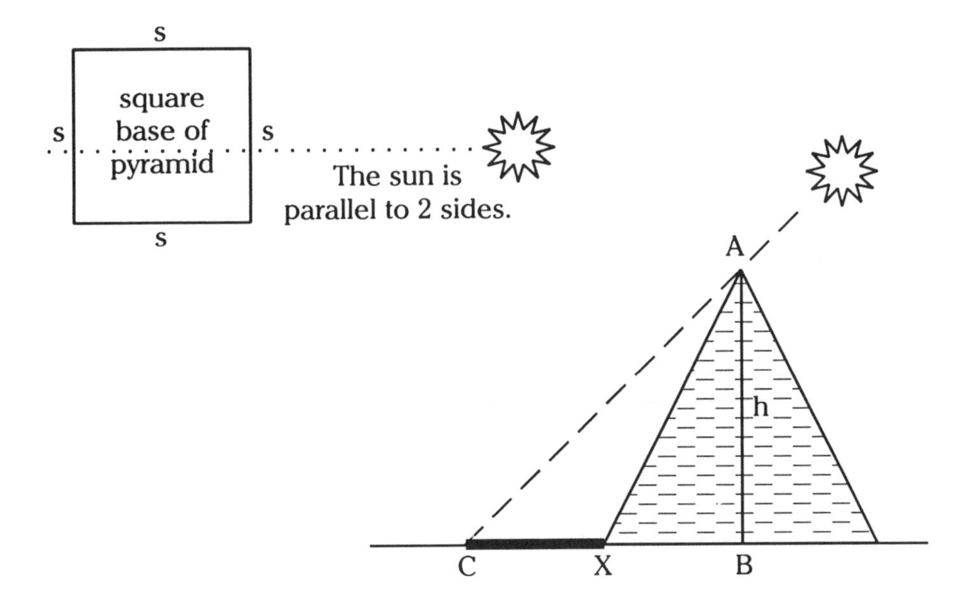

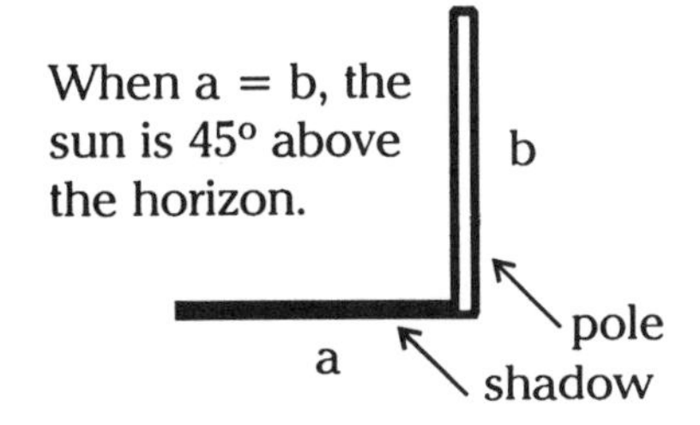

DOUBLING THE ANGLE ON THE BOW IN BOATING

This method uses geometry to determine the distance of a boat from a stationary object, by taking advantage of the fact that legs opposite equal angles (isosceles triangle) are equal. Two people are needed in the boat, the helmsman and someone to take bearings with a pelorus or accurate compass.

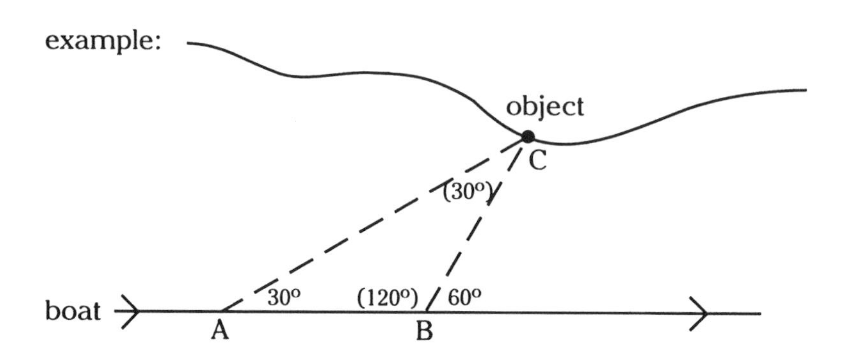

$\angle B = 2 \times \angle A$

AB (the distance the boat travels) = BC

DISTANCE/SPEED/TIME

Distance equals speed multiplied by the time: D = s x t
One may also use the word "rate" for speed: D = r x t

Problem: How far does a car travel at 65 mi./hr. for 2 hours?

D = s x t
D = 65 mi./hr. x 2 hours
D = 130 miles

Boating formula: 60 D (nautical miles) = s (knots) x t
This is called the "60 D Street" formula

$$D = \frac{s \times t}{60}$$ (in minutes) (in order to convert to mi./hr.)

Sir Isaac Newton (1642-1727), a professor of mathematics in England who counted among his accomplishments the invention of the calculus and the discovery of the law of universal gravitation, said that the force between bodies is directly proportional to the product of their masses and inversely proportional to the square of the distance between them. Acceleration of a falling object due to gravity at the earth's surface varies slightly because of various factors, but in common use is g (acceleration) = 32 ft./sec.^{2}.

ALGEBRA

Algebra is a most useful tool in everyday life. By using algebraic terms, thoughts and words are organized and simplified, and words become letters, numbers, or symbols. The beauty of logic takes over in the solutions when an equation is set up to find the roots or unknowns.

Procedures on both sides of the equation must be the same. If you multiply by a number on the left of the equal sign, you must do the same on the right side. This rule also applies to dividing, adding, and subtracting.

Example: $3x = 15$

$$\frac{3x}{3} = \frac{15}{3} \quad \text{(divided both sides by 3)}$$

$$x = 5$$

Problems:

Anne is twice as old as Bob. In 10 years, the sum of their ages will be 29 years. How old is each now? Let x = Bob, $2x$ = Anne

$(x+10) + (2x + 10) = 29$ (combine like terms)

$3x + 20 = 29$ (subtract 20 from both sides)

$3x = 9$ (divide both sides by 3)

$x = 3$ (Bob's age)

$2x = 6$ (Anne's age)

The difference between two numbers is 3. Their sum is 15.
Let x = the larger number, x-3 = the smaller number.

$x + (x-3)$	=	15	(combine like terms)
$2x - 3$	=	15	(add 3 on both sides)
$2x$	=	18	(divide both sides by 2)
x	=	9	
$x - 3$	=	6	

An easy method of making $1\ ^1/_2$ times a recipe is by using algebra. The original recipe calls for 1 C flour and $^3/_4$ C water.

$$\frac{3/4}{1} = \frac{w}{1\ ^1/_2}$$ (water) (change fractions to decimals)

$$\frac{.75}{1} = \frac{w}{1.5}$$ (cross multiply)

$w = .75 \times 1.5$ or 1.125 or $1\ ^1/_8$ C

Definitions:

Equation: a statement written in symbols. The side left of the equal sign is the same as the side right of the sign.

$6 + 4 = 10$ is a numerical equation
$x + 4 = 10$ is a literal equation because it has a letter (x).

Variable or "unknown": a symbol (a, b, c, x, y, etc.) which stands for a class of things.

Definitions (cont.)

Term: a member of a compound quantity

monomial:	$3xy$
binomial:	$(x + 3xy)$
trinomial :	$(x^2 + 3xy + y^2)$
polynomial:	an expression of two or more terms.

Similar terms can be combined: $3xy + 5xy = 8xy$

Degree: the sum of the exponents of the variables

The degree of 6 is 0, of $6x$ is 1, of $6x^2$ is 2, of $6x^2y^3$ is 5

The degree of a polynomial is the degree of the monomial term of the highest degree.

$3x^3 + 2x^2 - 3x + 1$: The degree is 3.

Coefficient: any numeral or literal symbol placed before another symbol or combinations of symbols as a multiplier.

$3x$: 3 is a coefficient of X.

"+" = a positive number; "-" = a negative number

Rules:

$+ \bullet + = +,\quad - \bullet - = +,\quad + \bullet - = -,\quad - \bullet - = +$

Those who took elementary algebra may remember:

Finding the roots or unknowns of a quadratic equation

$$\begin{aligned} x^2 - x - 6 &= 0 \\ (x+2)(x-3) &= 0 \\ x &= -2 \\ x &= 3 \end{aligned}$$

Multiply to check

$$\begin{array}{r} x + 2 \\ \underline{x - 3} \\ x^2 + 2x \qquad \\ \underline{\quad - 3x - 6} \\ x^2 - x - 6 \end{array}$$

Definitions (cont.)

Difference of two squares:

$(a^2 - b^2) = (a + b)(a - b)$ for any a and b

$144a^2 - 25b^2 = (12a + 5b)(12a - 5b)$

Linear Equation: The graph of a first degree equation which is a straight line.

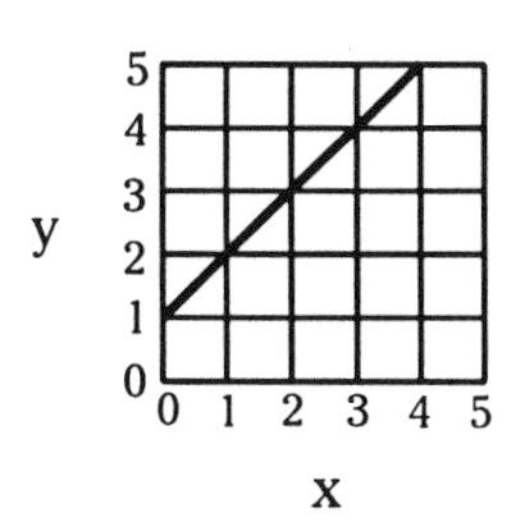

$y = x + 1$

Ratio: two quantities expressed as a fraction with the first quantity as numerator and the second quantity as denominator

Of the 100 people in the room, 75 are men

$$\frac{75}{100} = \frac{3}{4} = 75\% \text{ (men)}$$

Divide a 56" board into 3 lengths with a 1:2:4 ratio

$$\begin{aligned} 1x + 2x + 4x &= 56'' \\ 7x &= 56'' \\ x &= 8'',\ 2x = 16'',\ 4x = 32 \end{aligned}$$

Proportion: the equality of two ratios

$\frac{a}{b} = \frac{c}{d}$ or a:b::c:d (read a is to b as c is to d)

The product of the means (a & d) = the product of the extremes (b & c), so ad = bc.

MONEY

Interest formula: $I = prt$ (Interest = principal x rate x time)

Example: $I = \$1000 \times 5\% \times 1 \text{ year} = \$50.00$

Simple interest is computed annually on the principal.

Divide 72 by the rate to determine the number of years needed to double money:

e.g. $72 \div 8 = 9$ years

8% = 8 per 100 (per centum or by the hundred)

Compound interest is interest on the principal as well as on any previous interest added.

$$A \text{ (amount)} = p\left(1 + \frac{r}{n}\right)^{nt}$$

$p = \$5000$, $r = 0.09$, $t = 12$ months

"n" is the number of times it is compounded per year

e.g.: semiannually

$$A = \$5000\left(1 + \frac{0.09}{2}\right)^{2 \times 12}$$

$$A = \$5000\,(1 + 0.045)^{24}$$

$A = \$5000\,(1.045)^{24} =$ the number of times 1.045 must be used as a factor.

This is solved by logarithms or with a calculator.

Yield is the true return rather than "now money" (interest over cost). It may be higher or lower than the interest rate. If a $10,000 bond is bought at par at 8% interest, the yearly payment of $800 divided by the $10,000 gives a yield of 8%. If this bond is later bought at a discount rate for $8,000, the yearly payment of $800 gives a yield of 10%. If the bond is then bought at a premium $12,000, the yearly payment of $800 gives a yield of 6 2/3%.

Yield to maturity is determined by a complicated formula so you might like to leave it to the experts.

Money:

What is this stuff that makes us climb out of a warm bed each morning to face a cold world?

Well, not much actually. Those pieces of paper used to be notes promising to pay the bearer so much gold or silver, but no more. Now it just says it's worth so many dollars, nothing more, and doesn't promise to pay the bearer something else. Makes you nervous, doesn't it?

Good reason. In the past—and still, in countries plagued with high inflation—these notes came on bad times. Have you tried cashing in a Confederate bill lately?

So why do people accept a piece of paper as valuable exchange? Because they know that other will also. Act of faith. Simple, huh?

How did money come about? It's been around for so long, in so many various forms, that no one is sure, but Aristotle remarked on the need for it, saying, "The various necessities of life are not easily carried about..." He is right. It is tough to stuff a pig, chicken, or ear of corn in your wallet or purse to go shopping.

—JDJ

MECHANICAL ADVANTAGE

Law of the Lever

Archimedes (287-212 B.C.), a Greek mathematician and inventor who wrote a treatise on math which is still in existence, said that given a lever long enough, he could move the earth.

$$WL = W'L' \text{ or } \frac{W}{L'} = \frac{W'}{L}$$

W	=	weight to be raised
L	=	distance of weight from fulcrum
W'	=	effort exerted
L'	=	distance of effort from fulcrum

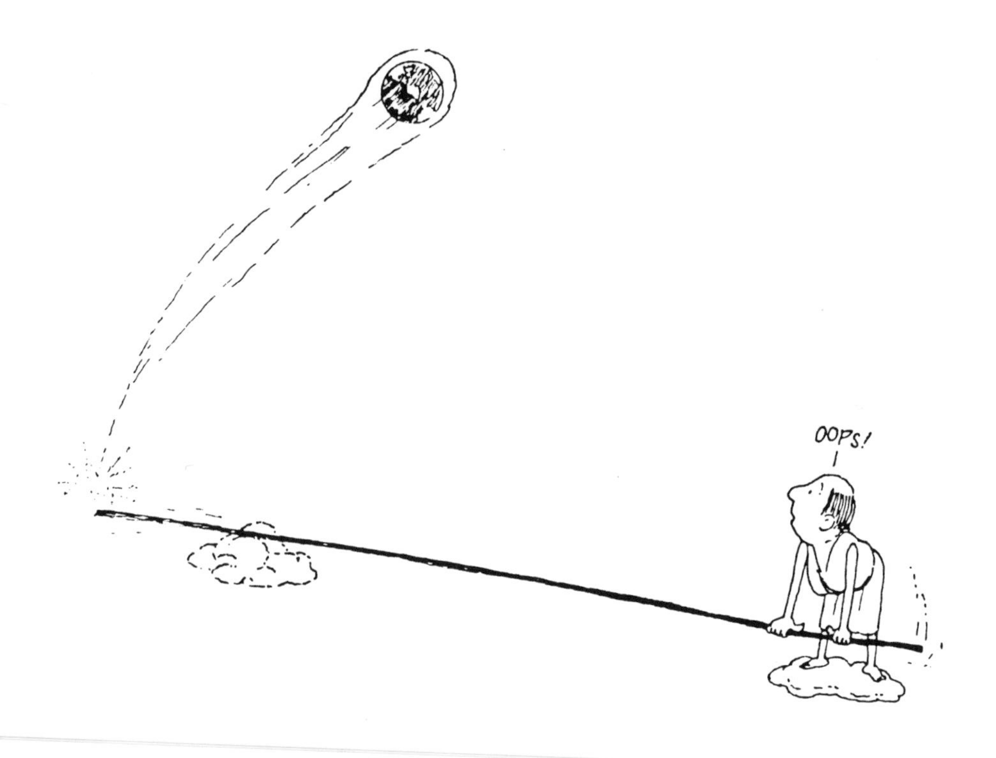

$$\frac{200 \text{ lbs}}{4 \text{ ft}} = \frac{x \text{ (effort)}}{2 \text{ ft}} \qquad \begin{aligned} 4x &= 400 \\ x &= 100 \text{ lbs effort} \end{aligned}$$

If you increase the distance of the effort from the fulcrum, less effort is needed.

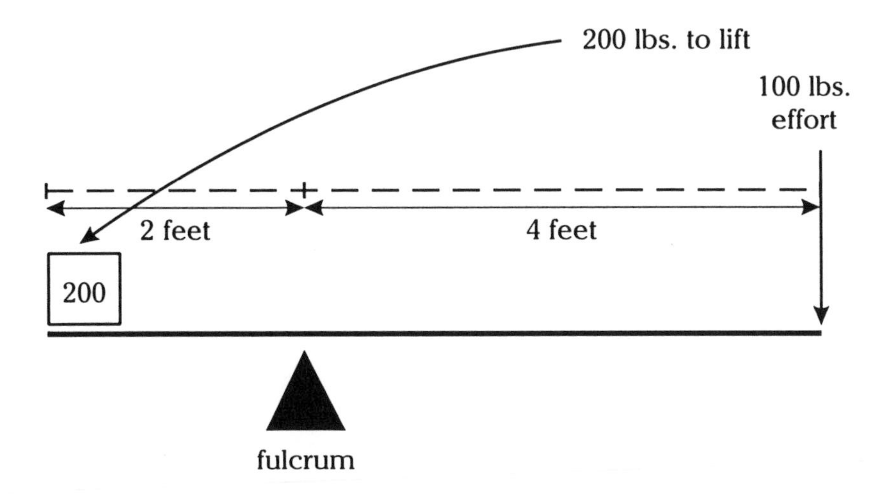

Children make proper adjustments for weight when using a seesaw.

MEASUREMENTS AND EQUIVALENTS

(English and Metric)

This listing, compiled for easy reference, gives only the most commonly used terms, unusual terms, measurements often found in literature, and specialized terms encountered in daily living. For further listings consult a dictionary, an encyclopedia, a math reference book, or an almanac.

Linear Measurements

click (klick, klik) = one kilometer (armed forces)

1 statute mile = 5280 feet or 1.609 kilometers

1 kilometer = .621 miles

1 meter = 39.37 inches

1 inch = 2.54 centimeters

1 rod = 5.5 yards

1 furlong = 220 yards

1 hand (horses) = 4 inches

1 league = 2.4 to 4.6 miles
according to the time and country;
about 3 miles in English speaking
countries; used poetically

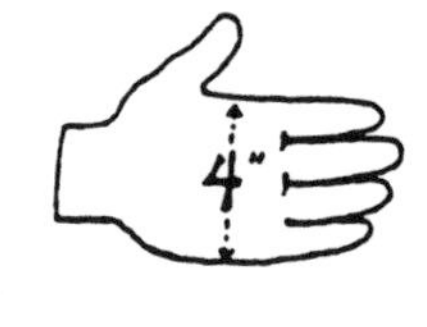

1 link = 7.92" of a surveyor's chain, 1 foot of an engineer's chain

1 light year = the distance light travels in a year
(about 6 million million miles)

Area and Volume

1 acre = 1/640 of a square mile; originally the area a yoke of oxen could plow in a day (varied)

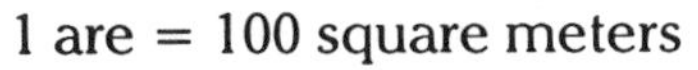

1 are = 100 square meters

1 hectare = 100 ares

1 section = 1 mile square (640 acres)

1 township = 36 sections or 36 square miles

1 cord = 128 cubic feet (firewood): 4 feet x 4 feet x 8 feet

1 lug = a 25 to 40 lb. box or basket of produce

Liquid (capacity and volume)

1 tablespoon = 3 teaspoons

1/2 cup = 1 gill (pronounced jill) (whipping cream)

8 tablespoons = 1/2 cup (1 cube of butter)

1 pint = 16 fluid ounces, or 2 cups

1 liter = 1.05+ quarts

1 quart = .95 liters

1 barrel (bbl) = 31 to 42 gallons

1 hogshead (a large cask) = 63 to 140 gallons

1 dram = 1/16 ounce

1 gallon = 4 quarts, or 8 pints, or 32 gills

1 imperial gallon = 1.2 U.S. gallons; (in some Commonwealth nations)

1 firkin = 1/4 of a barrel usually (There is reference to firkins in the King James version of the Bible, John 2:6)

Dry Capacity

1 peck = 2 gallons

1 bushel = 4 pecks

Avoirdupois Weight

1 short ton = 2000 pounds

1 long ton = 2240 pounds

1 kilogram (kilo) = 2.2 pounds;
$^1/_2$ kilo = 1.1 pounds

1 pound = .45+ kilograms

1 pound = 7000 grains

1 pound = 16 ounces

1 stone = 14 pounds (Great Britain)

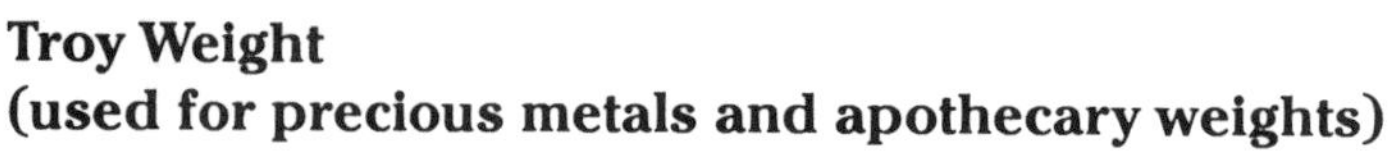

Troy Weight
(used for precious metals and apothecary weights)

1 pound = 12 ounces

1 pound = 5760 grains
(373 grams)

1 ounce = 20 pennyweights
or 480 grains (31.1 grams)

1 grain = 0.0648 grams

1 carat (karat) = 200 milligrams
or 3.086 grains

Nautical

1 fathom = 6 feet

1 nautical mile = 1.15 statute miles

1 knot = 1 nautical mile per hour

Circular (map or globe)

60 minutes (') = 1 degree (°)

60 seconds (") = 1 minute

Time

1 score = 20 years

1 millennium = 1000 years

1 fortnight = 14 days (2 weeks)

1 month: 30 days has September,
April, June, and November.
All the rest have 31, except February which has 28,
or 29 in leap year every four years.

Heat Capacity

Calorie (metric): The amount of heat needed to raise one gram of pure water one degree Celsius

British Thermal Unit (BTU): The amount of heat needed to raise one pound of pure water one degree Fahrenheit

Miscellaneous

Baker's Dozen = 13

Brace: a pair, a couple (ducks, pistols, etc.)

Chain = 100 links (The engineer's chain is 100 feet long or 100 links; the surveyor's chain is 60 feet long.)

Cloth measure: 1 bolt = 100 yards

Computer: 1 letter = 8 bits;
8 bits = 1 byte; 1 byte = 1 character;
1024 bytes (rounded off to 1000 metric) = 1 kilobyte

Hertz (Hz) = 1 cycle/second;
megahertz (Mhz) = 1 million Hz;
kilohertz = 1000 Hz;
microhertz = 1/1,000,000 Hz;
AC current in the USA is 60 hertz

Horsepower = 33,000 foot pounds/minute or 746 watts

Legion: 3000 to 6000 men in 8 ranks in the Roman army

Lumber: 1 board foot—1 foot square board, 1 inch thick

Milligauss: unit of magnetic energy density

Money: 1 mill = 1/10 of a cent

Paper: quire = 24 (or 25) sheets;
ream = 20 quires (480, 500, or 516 sheets);
case = 10 reams

Troika: Russian team of 3 horses abreast,
or a vehicle drawn by 3 horses

Typography: 1 point = 0.013837 inches
or 0.351 millimeters; 1 pica = 12 points

Yoke: 2 animals yoked together,
joined at the neck (oxen)

MEASUREMENT STANDARDS

The English System is older and less practical than the metric system. The U.S. System was inherited from the English, but is now different. The metric (decimal) system, based on multiples of ten, was originated by the French in the 1790s. This system eliminated difficult fractions and memorization. It is now the International System (Metric Standard located at the International Bureau of Weights and Measures at Sèvres near Paris). The United States doesn't fully accept the metric system because of the expense for industry, and for the public confusion which would result.

TERMS AND DEFINITIONS

Algorithm: any systematic procedure in math

Angles: acute—more than 0°, less than 90°
right—90°
obtuse—more than 90°, less than 180°
straight—180°
reflex—more than 180°, less than 360°
oblique—acute or obtuse angle
supplementary—a total of 180°
complementary—a total of 90°

Arc: a portion of a curved line

Area: surface measurement of a geometric figure

Arithmetic terms:
addend + addend = sum
minuend - subtrahend = difference
multiplier x multiplicand = product
dividend ÷ divisor = quotient

Axiom: a general statement we accept as true without proof (e.g., The whole is equal to the sum of its parts and is greater than any one of them).

Bisector of an angle: a line which meets the vertex of an angle and separates the angle into two equal parts

Bit: shortened term for binary digit

Byte: 8 bits

Coefficient: any numeral or literal symbol placed before another symbol as a multiplier ("2" in 2ab, "1"in x, "-4" in -4a)

Collinear: points lying on the same straight line

Concave: a surface that curves inward at the center; interior of a curved surface

Congruent: exact coincidence throughout; superimposable

Convex: a surface that curves outward at the center; exterior of a curved surface

Coordinates: any of a number of magnitudes that determine position, especially of spatial elements, as of point, planes, etc.

Coplanar: points lying on the same plane

Corollary: a theorem derived from another theorem or postulate with little or no proof

Decimal: proceeding by tens, each unit being ten times the unit next smaller (e.g., 0.102)

Decimal fraction: denominator is some power of ten (e.g., $0.2 = \frac{2}{10}$)

Denominator: the number below the line in a fraction indicating into how many equal parts the unit is divided

Equivalent: equal in value but not admitting of superposition

Exponent: power of the number (e.g., x^2: x = base, 2 = exponent)
Add exponents in multiplications ($10 \times 10 \times 10^2 = 10^4$)
Subtract exponents in division (e.g., $x^5 \div x^3 = x^2$)

Factoring: the opposite of multiplying;
3 is common factor of 6 and 9

Prime numbers: 2, 3, 5, 7, 11, 13, etc. (no "0" or "1"); numbers which cannot be factored

Every positive integer is or is not a prime number

Fraction: one integer divided by another
(See numerator and denominator)

Fulcrum: a pivot supporting a device such as a lever so it can balance

Geodesic dome: triangular facets roughly a hemisphere used in construction, for strength, by Buckminster Fuller

Geometric figures:

Circle: a plane closed curve, all points of which are equidistant from the center

Cone: bottom is a circle, sides taper up to a point

Cube: regular solid of 6 equal square sides

Cylinder: the surface traced by one side of a rectangle rotated around the parallel side as axis

Ellipse: a plane curve, whose distances from two fixed points is constant

Hemisphere: a half sphere, separated from a sphere by a plane through its center

Line:

vertical | horizontal | oblique

straight | broken | curved

Line segment: part of a line

Geometric figures (cont.)

Parallelogram: a quadrilateral with opposite sides parallel and equal

Pentagon: a polygon with five sides and five angles

Polygon: a figure generally plane and closed, having many angles and straight line sides

Prism: a solid whose bases are similar, equal, and parallel polygons, the faces being parallelograms

Quadrilateral: a plane figure of four straight sides and four angles

Rectangle: a parallelogram with right angles

Rhombus: an equilateral parallelogram with its angles oblique, and with two equal adjacent sides

Sphere: a body of space bounded by one surface, all points of which are equally distant from the center

Square: a parallelogram with equal sides and four right angles

Simplex:

• = 0-dimensional simplex

/ = 1-dimensional simplex

△ = 2-dimensional simplex

◭ = 3-dimensional simplex

Trapezoid: a plane 4-sided figure with two parallel straight sides

Triangle: a plane 3-straight sided figure

equilateral triangle:	3 equal sides
isosceles triangle :	2 equal nonparallel sides
right triangle:	1 right angle
scalene triangle:	no 2 sides congruent

Googol: 1 with 100 zeros after it

Hypotenuse: the side of a right angled triangle that is opposite the right angle. It's the shortest way to cross a rectangular vacant lot

Integer: a whole (not fractional or mixed) number

Locus: a figure formed by all the points which satisfy certain conditions; the path of a point or curve moving according to the same law (plural: loci, pronounced losī)

Logarithm (log): A logarithm of a number is defined with respect to a base of logarithms. The relation between a number N and its log (to the base) L is $N = B^L$ (B = base; L = log of N). The most common base is 10. Logarithms with base 10 are called common logarithms. e.g., $1000 = 10^3 = 10 \times 10 \times 10$. The log (to the base 10) of 1000 is 3. The base 2 is useful in dealing with binary numbers.

Numbers:

Ordinal: indicates order or succession (first, second, third, etc.)

Cardinal: used in simple counting (1, 2, 3, etc.)

Rational: ratio of 2 integers $15\ (\frac{15}{1})$, $\frac{5}{8}$, $2\frac{1}{2}\ (\frac{5}{2})$, 0.6666

Irrational: not expressed as an integer or as the quotient of 2 integers; 2; root of an algebraic equation

Transcendental: can't be the root of an algebraic equation; pi (π)

Real: union of the set of rational numbers and the set of irrational numbers; representable by an infinite decimal expansion, which may be repeating or non-repeating, and which includes zero and transcendental numbers

Numbers (cont.)

Prime: can be divided by no other whole number than itself or 1 (e.g., 2, 5, 13, 89, etc.)

Composite: can be divided by some number other than itself

Even: divisible by 2

Odd: has remainder of 1 when divided by 2

Number Line: negative positive

-2 -1 0 +1 +2

Numerator: the number above the line in a fraction

Palindrome: 1221; 33233; 45654

Parallel Lines: 2 lines on the same plane which don't intersect

Perimeter: the whole outer boundary of a figure or the measure of the same; the sum of the lengths of the sides of a polygon

Pi (π): ratio of the circumference of a circle to its diameter; 3.14285714..., 3.14 (used but not accurate), 3 1/7, 22/7. The approximate value of pi was calculated by Archimedes

Postulate: a proposition which is taken for granted or put forth as axiomatic; geometric statement which is accepted as true without proof

Powers: 3562.143:

$$(3x10^{3})+(5x10^{2})+(6x10^{1})+(2x10^{0})+(1x10^{-1})+(4x10^{-2})+(3x10^{-3})$$

Radius: a line from the center of a circle or sphere to the curve or surface

Ray: the part of a straight line extending infinitely far in one direction from a point

Reasoning:

Intuitive: no scientific observation or experimentation

Deductive (proof): If all the information is true, then the conclusion is true

Inductive: probable answer only (looking for a pattern); to arrive at the truth by investigating a number of cases

Reciprocal: The product of a number and its reciprocal is 1 or unity. (e.g., 4/1 is the reciprocal of 1/4; 3/4 is the reciprocal of 4/3)

Set: a number of things of the same kind that belong together

Empty Set: none (prime numbers over 3 divisible by 2)

Skew Lines: two lines in space neither parallel nor intersecting (not in the same plane)

Similar: figures of the same shape (not the same size)

Theorem: a statement to be proved

Topology: "rubber sheet geometry"—Figures are equivalent, not congruent

Transversal: a line which intersects two or more lines

Triangles: (See geometric figures)

Vector: directed line segment (speed and direction)

Vertex: the point where two lines defining an angle meet or points where two straight line segments of a polygon meet

SUBJECTS FOR POSSIBLE EXPLORATION

- Branches of mathematics: Geometry, Algebra, Arithmetic, Trigonometry, Applied Math, Probability, Statistics, Calculus
- The development of math for practical needs: agriculture, business, navigation, surveying, industry
- The spread of math developments by merchants
- Countries and peoples involved in the development of math: Egypt, Mayans, Mesopotamia, Aztecs, India, China, Persia, Greece, Arabs, Chaldeans
- Mathematicians
  - —The three greatest from the past: Archimedes, Gauss, Newton
  - —Other notable mathematicians include: Thales, Pythagoras, Plato, Aristotle, Hypatia, Euclid, Ptolemy, Archimedes, Copernicus, Galileo, Kepler, Descartes, Leibnitz, Pascal, Euler

  - —Some of the greatest mathematicians of the 19th and 20th centuries: Cantor, Poincaré, Hilbert, de la Vallée Poussin, Cartan, Levi-Civita, Lifschetz, Courant, Seigel, Gödel, Mandelbrot, and von Neumann
  - —John von Neumann (1903-1957), probably the best known of recent contributors to mathematics, was affiliated with the Princeton Institute for Advanced Study after 1933. His work on the atomic bomb, and on computers was significant.

TRIGONOMETRY
CALCULUS
STATISTICS
ALGEBRA
APPLIED MATH
GEOMETRY
METIC
ARI
INTEGRAL
PROBABILITY

BIBLIOGRAPHY AND SOURCES

The New Columbia Encyclopedia (New York: Columbia University Press, 1975) 3052 pp.

Encyclopedia Americana (International Edition; Danbury, Connecticut: Grolier, Inc., 1990) 30 vols.

Apel, *Willi. Harvard Dictionary of Music* (Cambridge, Massachusetts: Harvard University Press, 1964) 833 pp.

Webster's Collegiate Dictionary (5th ed.; Springfield, Massachusetts: C. and C. Merriam Co., 1947) 1274 pp.

World Almanac (New York: Pharos Books, 1990) 960 pp.

King James Version, *The Old Testament* and *The New Testament* (New York: Harper and Bros.)

Abbott, E.A. *Flatland* (New York: Barnes and Noble, Inc., 1965) 108 pp.

Garland, Trudi Hammel. *Fascinating Fibonaccis* (Palo Alto, California: Dale Seymour Publications, 1987) 103 pp.

Scholes, Percy. *Oxford Companion To Music* (5th edition; London: Oxford University Press, 1965) 1195 pp.

Class notes

Friends

Teachers

Sources lost in memory

INDEX